Re-issue of December Entangled Mag
Key Findings

The December issue of Entangled magazine represents the first time since its founding in June of 2017 a second edition was found necessary to publish.

New revelations became apparent shortly after the December 1st publication. Specifically, Ezekiel's wheel within a wheel and its direct correlation to the design and operation of the particle accelerators and colliders of the LHC at CERN. And, the use of the detectors, ALICE, ATLAS, CMS, LHCb AND LHCf as an early warning system of spiritual energies and entities.

Included is a breakdown of the physics involving the use of Strangelets for the opening of the doorway to the Abyss.

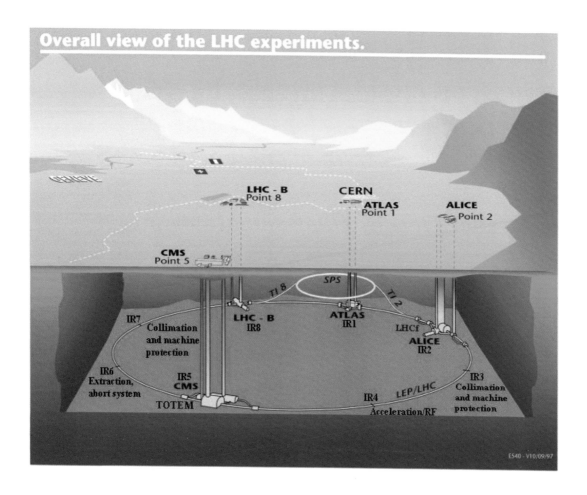

In 2009, Dr. Sergio Bertolucci, former Director for Research and Scientific Computing of the facility, now famously stated the Large Hadron Collider could for *"a very tiny lapse of time to peer into this open door, either by getting something out of it or sending something into it."*

"Of course," added Bertolucci, *"after this tiny moment the door would again shut; bringing us back to our 'normal' four-dimensional world ... It would be a major leap in our vision of nature... And of course [there would be] no risk to the stability of our world."*

His follow on statement:

Dr. Bertolucci later confirmed that yes indeed, there would be an "open door", but that even with the power of the LHC at his disposal (in 2009) he would only be able to hold it open *"a very tiny lapse of time, 10-26 seconds, [but] during that infinitesimal amount of time we would be able to peer into this open door, either by getting something out of it or sending something into it."*

▶ Symmetry

https://vimeo.com/120676848

CERN produced video: Symmetry (trailer).

ALICE (A Large Ion Collider Experiment) is a heavy ion detector, for collecting data from the head-on collisions of heavy ions of lead (Pb to Pb), producing the quark-gluon condensate known as Strangelets. This is a choreographed scene taken from the CERN produced video, Symmetry, known as the "Dance of Death", performed in front of the ALICE detector. Strangelets are the most explosive substance in the known universe during the primordial moments at the creation of matter.

This production is currently underway until December 10, 2018.

Circle scene from Symmetry, as used in Black Magic rituals. Energy is contained within a circle. The figure in white is dancing in a counter-clockwise direction, depicting several things.

One, the inherent spin of quantum particles.

Two, one of the two contra-rotating (two) beams of accelerated particles within the Main Ring of the Large Hadron Collider.

Three, the figure in black, is Satan.

There is a great deal more contained and signified within their self-professed Symmetry video. The focus here is the use of Strangelets, the most powerful explosive substance in the known Universe, in opening of a doorway, a portal, to another dimension. Again, self-professed. Herein, I will explain the physics involved in this opening of the doorway to the abyss. Please refer to the many articles previously published in this magazine, dating back to June of 2017. I have been writing and broadcasting about CERN since 2011, citing what I am presenting to you now, in December of 2018, and much more. The Large Hadron Collider at CERN is scheduled to shut down, for Long Shutdown 2 on December 10, 2018. Or will it?

This image, taken from Symmetry, depicts their own admission of an explosion resulting in the opening of the inter-dimensional doorway to the abyss. Please refer to the 1 min 16 sec point in the trailer of this video for a closeup image of the control center instruments indicating "Collision: Failed".

https://vimeo.com/120676848

Please refer also to the 14 sec point where the sacred geometry employed at CERN is displayed. God created the geometry of the universe. The enemy counterfeits God's creation, including mathematics. Of note, is the tetrahedron as the first figure of the Platonic solids in the video. Please refer to my work regarding the tetrahedron within past issues of this magazine.

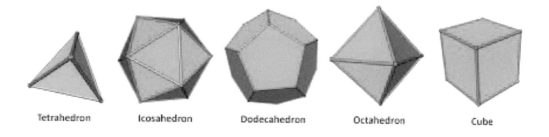

Tetrahedron   Icosahedron   Dodecahedron   Octahedron   Cube

And now for the Strangelets themselves. The material employed in the actual opening of the abyss as depicted within Revelation 9.

https://www.biblegateway.com/passage/?search=Revelation+9&version=KJV

Rev. 9

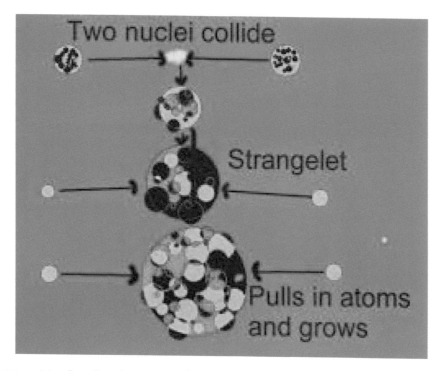

Two nuclei collide

Strangelet

Pulls in atoms
and grows

http://adsabs.harvard.edu/abs/1996APS..DNP..FF09D

Abstract from 1996 white paper:  Strangelets

Experiment E864 at the BNL (Brookhaven National Laboratory) AGS is searching for neutral strangelets using an 11.6 A GeV/c Au beam impinging upon a fixed target of Pb. An 18 quark neutral strangelet, the Quark-Alpha, constituting a spin, color, and flavor singlet might be particularly stable, and thus, susceptible to detection. The detector is a spectrometer with a large acceptance about mid-rapidity followed by a hadron calorimeter. The energy and time of flight of neutral particles are measured by the calorimeter. For neutral particle searches, the tracking planes of the spectrometer provide charge particle rejection. A "late energy trigger" (LET), which selects events with a large deposit of energy in the calorimeter coming late in time, preferentially chooses events with massive particles. The 1995 data set is the first to be collected with the completed detector and LET. The current status of the search for neutral Strangelets in the 1995 data will be discussed.

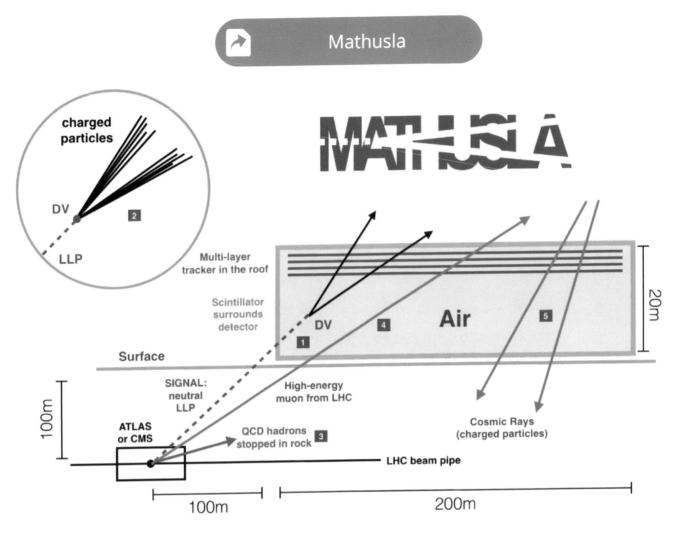

MATHUSLA, (Massive Timing Hodoscope for Ultra Stable Neutral Particles), named after the longest-living man in the Book of Genesis. Its job: to hunt for long-lived particles that the LHC cannot detect itself. A hodoscope is a device capable of detecting the paths of subatomic particles.

This new detector will be installed during the planned Long Shutdown 2 of the Large Hadron Collider, resuming experiments scheduled in 2026.

Again, current detectors such ALICE wherein Strangelets are produced during the collisions of heavy ions of lead (Pb to Pb), cannot detect the presence of Strangelets beyond only a few moments, then described by scientists as "winking out of existence". The reality of this is the fact both BNL and CERN have hard evidence that Strangelets are produced upon achieving an energy threshold of 10 TeV, a level both labs have produced since 1995 at BNL and since 2015 at CERN, and continuing today.

As the number of particles increase, most normal quarks convert into Strangelets in an exponential growth rate.

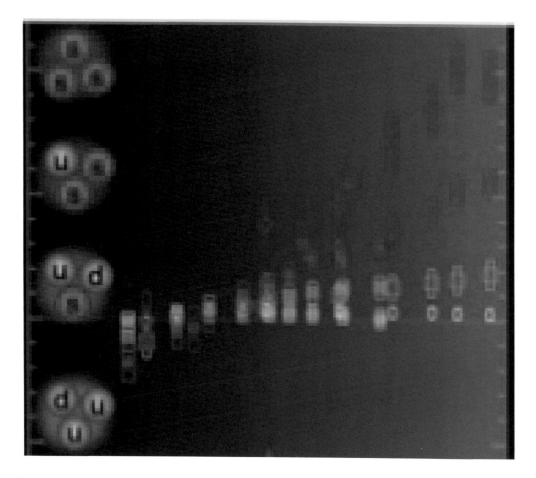

 Strangelets begin life as neutral dibaryons, a type of composite subatomic particle containing an odd number of valence quarks (at least 3). Baryons belong to the hadron family of particles, which are the quark-based particles.

Again, the present detectors installed along the beamline, the Main Ring of the LHC, are designed to measure the charged particles, or the product of the collisions which they themselves possess a charge. Strangelets are neutral, without a charge.

Three such dibaryons form a alpha MACHO (MAssive Cold HalO object) minimal Strangelet.

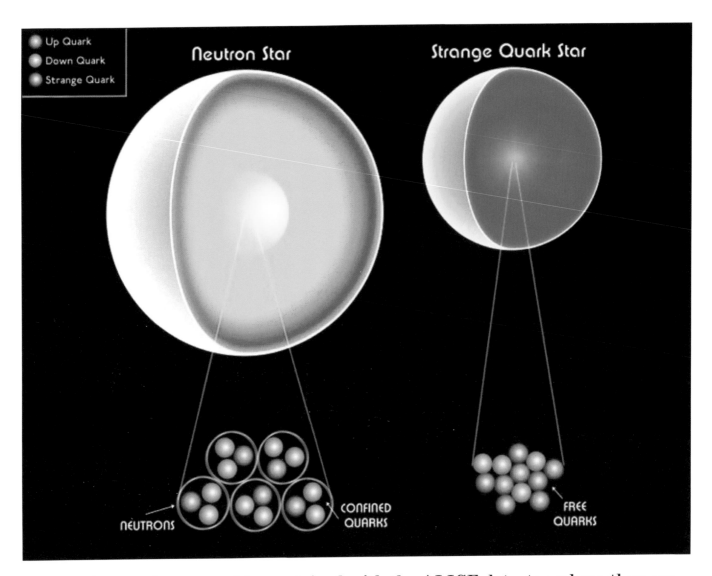

Neutral Strangelets cannot be contained with the **ALICE** detector where they are produced by collisions numbering in the millions of heavy ions of lead, thus free-falling to the center of our Earth. Along the way and once there, attracting all other matter to them.

The danger in this scenario resides in the collapse of the planet to some 15km in diameter, forming a neutron, or quark star. The question then is, within what time frame might this occur? It is a matter of how many Strangelets already have been produced both at **BNL** and **CERN**, and are presently.

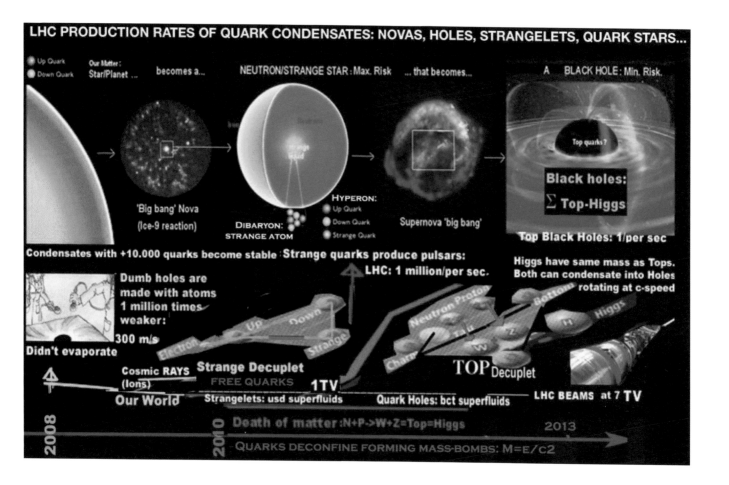

The alpha quark MACHO (MAssive Cold HalO object) is the minimal unit of QSM (Quark Strange Matter) and all the other particles of the chain are extremely stable, as all systems of nature have ternary patterns of growth, from genes to scales of nature. This ternary pattern is reflected in the tetrahedron, the simplest polytope enclosing a volume, seen in graphene cells of C_{60} carbon, the cells of honey bee combs, to tree cells.

Ternary (composed of 3 parts) follows a line of progression to reach a crystalline state of strange quark stars in a ternary progression of hyperons (usd)->(3 dibaryons: 18 quarks)->

Pulsar "Pulsating Star"

To answer the question of when there will be enough triplets of dibaryons (alpha quark MACHOS) to start the collapse of the Earth...a few thousand will be enough.

Given Strangelets are neutral, they will require gravity, not electromagnetism, to gather at the center of the Earth as a 'gravitational lump', occupying a small 10^{18} nucleus space. This forms a point of gravitational force the size of an atom, enough to begin the collapsing of the planet given the large density of the core.

These 'reproductive radiations' are extremely fast, growing exponentially, and is why Novas are so fast in the creation of a strange pulsar star.

Article: *Nuclear Detectives Hunt Invisible Particles That Escaped World's Largest Atom Smasher.*

A few observations are in order here:

1. At least a science-based publication, in this case, Live Science has been shall I say; "allowed" by CERN to publicly, at least to those reading of science, to acknowledge the production of sub-atomic quantum particles of which the Large Hadron Collider is ill-equipped to confine within its multi-story, multi-ton detectors. Specifically in this case, ALICE.

2. The LHC is *not* an "atom smasher". For a "science magazine", they should check their physics at CERN's doorway to the abyss. Sub-atomic particles, on the quantum scale, constituent *parts* of atoms, such as protons and heavy ions of lead and gold are *collided* together within the detectors of the LHC. I suppose one could assign the word 'smasher' to it if one really wanted to be old-fashioned in their choice of vernacular as carryover from the 1950s and 60s.

3. As to these particles themselves being "invisible", that is true enough given the fact the present detectors, including ALICE, are not designed to measure and record the particles thus realized from the current run of collisions involving heavy ions of lead. These are the quark-gluon condensates: Strangelets, of which, the detectors effectively are blind to, thus "invisible".

4. Lastly, to the naming of said particles realized from collisions of lead, albeit in smaller quantities via collisions of protons-to-protons, and protons with lead are produced...this article identifies these as "gluinos" and "composite dark glueballs". Oh come on now. To whom are they addressing as a supposed scientific magazine? These are incorrect names, and mislabeled as "speculative particles". Only because, and again, the present detectors cannot measure what is being produced, and *have* been confirmed mathematically as well as empirically, Strangelets - confirmed at both Brookhaven National Laboratory (BNL) using their Relativistic Heavy Ion Collider (RHIC) at 10 TeV and beyond, as yes, at CERN itself. How can these two machines measure the production of Strangelets? The detectors *are* but for a few brief picoseconds able to observe them, describing this event as "particles winking out of existence". This is deliberate obfuscation of Strangelets, akin to telling only half the truth.

5. The acknowledgement of the building of a new $50 million detector, METHUSLA, is disclosing two things:

A. A new type of detector is required; one designed specifically to detect then measure particles the present detectors are currently blind to.

B. By default, the acknowledged need for such a new detector confirms the past and present production of Strangelets within the Large Hadron Collider (LHC). Strangelets are produced at a threshold of 10 TeV. The LHC has realized 13.5 TeV during Run 2.

The machine is a quark-gluon condensate factory, manufacturing the most explosive substance in the known universe. This gave rise to the alleged Higgs boson, its discovery criticized and rightfully so in this article.

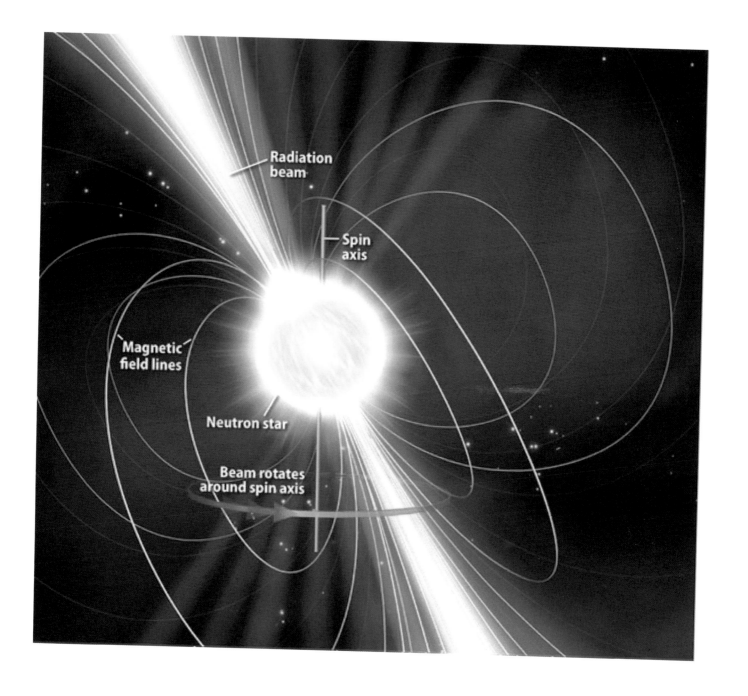

2 Peter 3:10 KJV

But the day of the Lord will come as a thief in the night; in the which the heavens shall pass away with a great noise, and the elements shall melt with fervent heat, the earth also and the works that are therein shall be burned up.

COVERT CATASTROPHE

BY ANTHONY PATCH

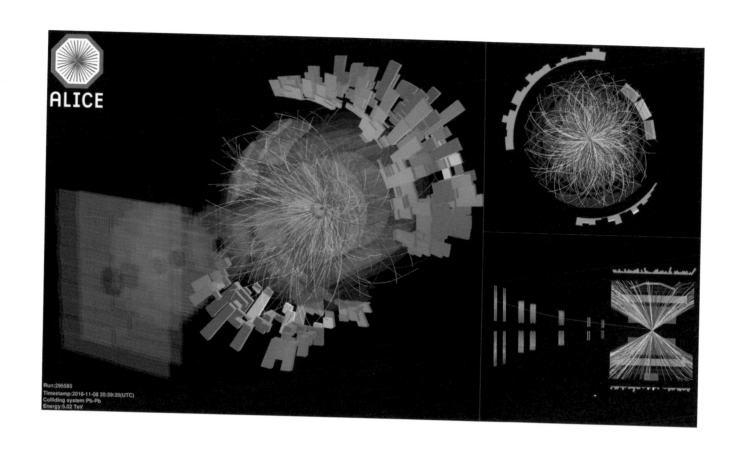

https://videos.cern.ch/record/1309872

https://videos.cern.ch/record/1702939

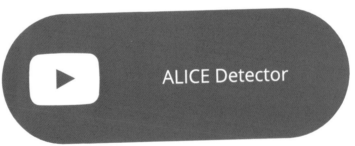

https://videos.cern.ch/record/2245570

Production of the quark - gluon condensate:

Strangelets.

https://www.higgypop.com/news/scientists-say-there-is-no-life-after-death/

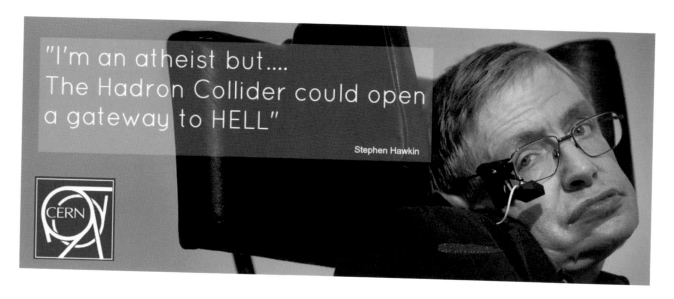

As I have so often stated, both on the radio and here in print, "you cannot separate the spiritual from the physical".

I have also urged listeners and readers seeking the truth, to "one-eighty it" when hearing or reading statements made by the media, and the scientific community. The late Dr. Stephen Hawking had made several statements admonishing those involved at CERN, warning them their activities with the LHC will lead to the destruction of mankind. No need to "180" his statement however.

The above article is, however, a prime example of when to apply my "180" method, in turning around statements, thus revealing the truth.

Recently, during several of the Livestreams with those who've hired me as their researcher and reporter through Patreon, I've revealed something more about the detectors of the LHC.

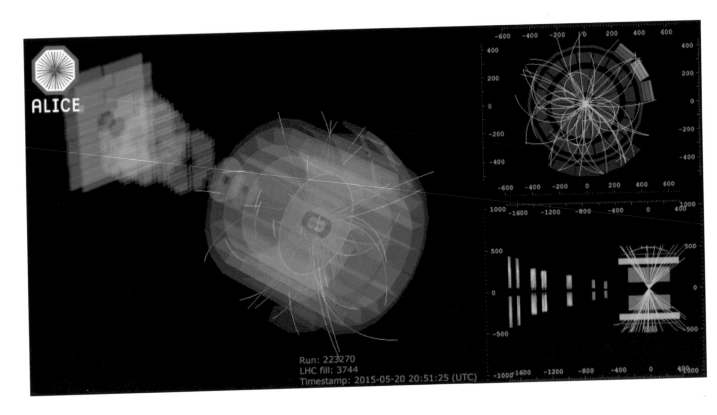

Examine the above graphic and the previous image depicting data received from collisions taking place within the ALICE detector. It is representative of the spiritual energies present as demonstrated by the Double Slit Experiment and the concept of Wave Particle Duality

https://en.wikipedia.org/wiki/Double-slit_experiment

https://en.wikipedia.org/wiki/Wave%E2%80%93particle_duality

Matthew 10:26 KJV

Fear them not therefore: for there is nothing covered, that shall not be revealed; and hid, that shall not be known.

I am sharing some of the media attention paid to science in a comparative manner deliberate in the application of my "180" method of seeing and conveying the obfuscated truth of the matter...or should I say wave...or both. You decide which, if indeed we are able to make such a distinction. A little levity goes light years.

Psalm 2:4 KJV

He that sitteth in the heavens shall laugh: the Lord shall have them in derision.

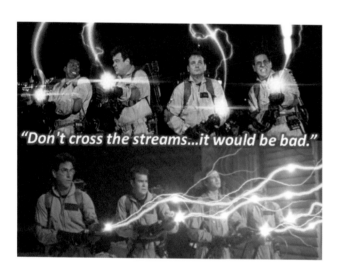

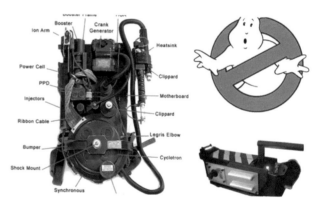

Note the Cyclotron, the large disc at the base of the pack, a proton particle accelerator.

The 1984 comedy film, Ghostbusters was a lot of fun, and much has been made of the spiritual and technological aspects of the plot and imagery. Presented here to demonstrate the cross-over and connection between the spiritual and physical. Although a movie, it does have specific relevance to serious ongoing research into the detection and proof as to the existence of spiritual energies and entities, especially at CERN.

"When we die our bodies decay but the particles that make us up aren't destroyed, they remain part of their respective quantum fields. So, in order for our consciousness to survived after death,it would need to exist as a quantum field of its own, but Dr. Carroll says that if that were the case we should be able to detect these "spirit particles" and "spirit forces." - **No Life After Death; Higgypop, May 2018**

Professor Brian Cox recently made a similar claim, stating that if some kind of spiritual energy existed the experiments at CERN (the European Center for Nuclear Research) should have detected it using the Large Hadron Collider (LHC), the world's biggest and most complex science experiment, a particle accelerator 17 miles in circumference built underground on the border of France and Switzerland. - **No Life After Death; Higgypop, May 2018**

No Life After Death

Now, back to the detectors of the LHC. Following your perusal of the above links (Double Slit Experiment and Wave Particle Duality) you see why I say; 'one cannot separate the spiritual from the physical'. For spiritual is comprised of energy. Something CERN's present and yet-to-be-upgraded detectors, including METHUSLA (a Biblical patriarch, who lived to 969 years of age, hence the reference to the real and actual longevity of Strangelets) will be enhanced for - detecting the waves of spiritual energy.

Akin to a military early warning detection system of missile launches, or approaching aircraft, so too are these detectors. Two groups of spirits are in play here: fallen angels, demons, and Satan, and God's angels, Jesus Christ, and the Holy Spirit.

Anticipating the ultimate battle ahead known as Armageddon, Satan seeks the mechanism to detect, apprehend, and annihilate God's Holy Spirit, Jesus Christ, and the entire angelic realm. Not to mention, keeping in line his own minions - for there will be great infighting amongst Satan's ranks.

Despite what scientists proclaim as to their complete knowledge of physics, etc., etc., they know nothing of the mind of God, therefore, of His eternal plan for His creation.

Presently, there is a peace plan due as this issue of Entangled magazine 'goes to press', to be announced by President Trump, he billing it as "the deal of the century". This is the planned joining of Gaza to the West Bank, thus dividing Israel.

These events are parallel to the scheduled shut down on December 10, 2018, of the LHC for the upgrading of the superconducting magnets, detectors, and many other major systems of the collider. Will those in control of the LHC attempt during this time frame an opening of the bottomless pit (Revelation 9:1-6)? I expect such an event to take place later in the 7-year Tribulation, in which God cuts short the days in order to preserve the human race. In 7 years (2026) the LHC is scheduled to begin collisions.

Matthew 24:22 KJV

And except those days should be shortened, there should no flesh be saved: but for the elect's sake those days shall be shortened.

Returning to our examination of Strangelets, I would like to offer my conclusion as to the particle physics, the quantum mechanics, even the quantum computing to be employed in the fulfillment of Revelation 9:1-6.

Patreon supporters benefit from their participation in private conversations held via Livestreams every Wednesday and Thursday at Noon EST. Much of what I've outlined here was discussed in depth during these live broadcasts. Livestreams, coupled with Entangled magazine, are vital learning tools.

During the Thursday, November 29, 2018 Livestream, I presented the following explanation as to how the 'inter-dimensional doors', or 'door' will be opened. This offered an alternative to the LHC fulfilling this role, citing perhaps this door will open within the confines of the black, cubical enclosure of one of D-Wave Systems adiabatic quantum computers. Below I will outline new insights.

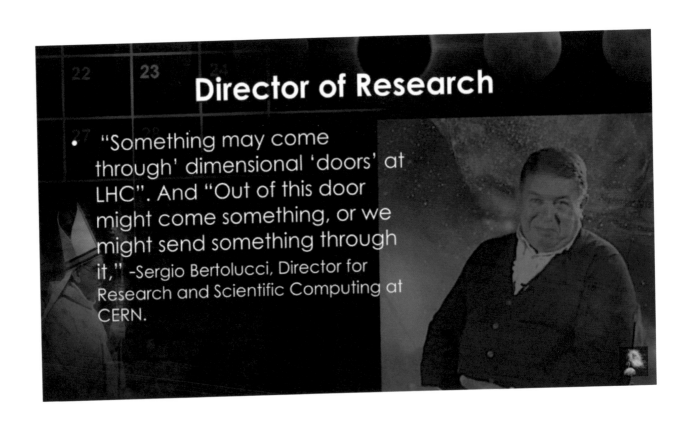

In the November 2017 issue of Entangled magazine, I detailed the quantum computer specifically, its chipset composed of qubits (quantum bits) operates in the same manner as that of the Main Ring of the LHC. This makes the particle accelerator/collider a 27km long quantum computer.

I believe the LHC will be used to open the door to the abyss. I believe they will use this machine as a weapon to detect, apprehend, and try to annihilate God, His Holy Spirit, Jesus, and the angelic realm. I have made a case for the quantum computer in performing as the controller of this event.

For if one is to begin opening this door, it must occur at the sub-atomic quantum particle scale. This brings quarks into play.

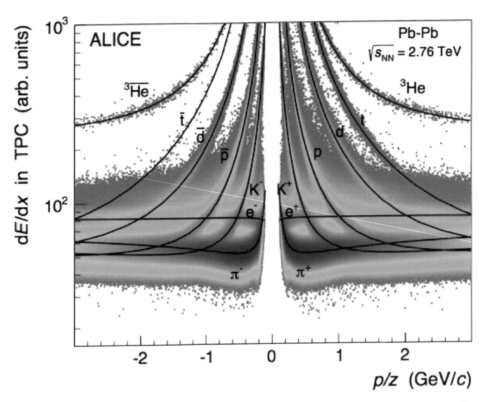

The above graph will function as an example of the process I am about to layout. It is not the actual 'doorway' event itself rather, something more tangible than simply talking about quarks and energies.

It is illustrating a collision event occurring within the ALICE detector, involving heavy ions of lead, the Pb-Pb seen in the upper right corner. Pb is the notation for lead taken from the periodic table of elements.

In the lower right corner, you will note (GeV/c) translated Giga (billion) electron Volts with /c as the speed of light.

What I offer to you is the white area running vertically between the multi-colored regions of the graph. I will take the liberty of demonstrating the separation of quarks, bound both by gluons, and what is known in particle physics as the Strong Force, or Strong Nuclear Force.

Fundamental Force Particles

Force	Particles Experiencing	Force Carrier Particle	Range	Relative Strength*
Gravity acts between objects with mass	all particles with mass	graviton (not yet observed)	infinity	much weaker
Weak Force governs particle decay	quarks and leptons	W^+, W^-, Z^0 (W and Z)	short range	
Electromagnetism acts between electrically charged particles	electrically charged	γ (photon)	infinity	
Strong Force** binds quarks together	quarks and gluons	g (gluon)	short range	much stronger

CERN will employ the LHC in the use of the most powerful explosive substance in the known universe, Strangelets, a quark-gluon condensate, to overpower this Strong Force, and the corresponding Force Carrier Particle, the gluon, thus separating other Strangelets. This separation is illustrated as the vertical white space in this graph.

Revelation 9:1 KJV

And the fifth angel sounded, and I saw a star fall from heaven unto the earth: and to him was given the key of the bottomless pit.

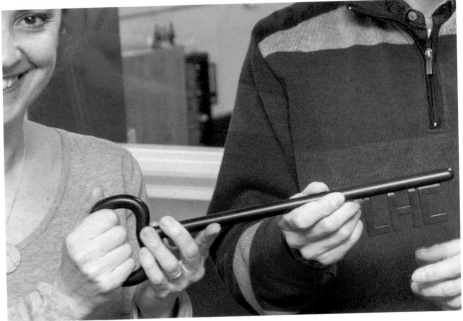

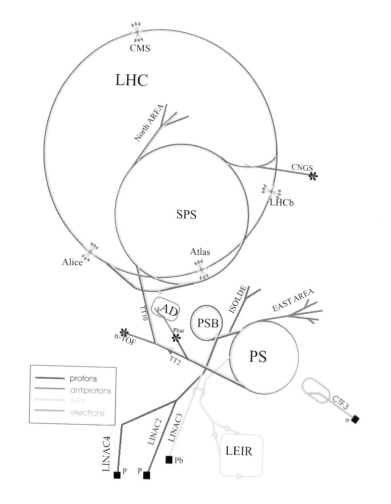

Within the center area of the Main Ring, the largest of the circles in this illustration of the linked accelerators of the LHC, is the North Area. This is an impact point, at the end of a straight beam line, a pipe carrying accelerated particles, to impact a barrier at the end of this beam line.

Normally, routine experiments are in this manner carried out at CERN, as well as all other Synchroton particle accelerators around the world.

Depicted here are a series of linked accelerators beginning with the LINAC linear accelerators (2,3,4) seen at the lower portion of this diagram. From here, accelerated protons are injected into the Proton Synchrotron Booster (PSB), then on into the Proton Synchrotron (PS) for increased acceleration. Next, into the Super Proton Synchrotron for additional acceleration and final injection into the Large Hadron Collider (LHC) for acceleration approaching the speed of light.

Proton Synchrotron, PS in the diagram from the previous page of the LHC.

Artist's concept of Ezekiel's wheels.

Ezekiel 1:16 KJV

16 The appearance of the wheels and their work was like unto the colour of a beryl: and they four had one likeness: and their appearance and their work was as it were a wheel in the middle of a wheel.

Ezekiel 1:20 KJV

20 Whithersoever the spirit was to go, they went, thither *was their* spirit to go; and the wheels were lifted up over against them: for the spirit of the living creature *was* in the wheels.

The Spirit In The Wheel

In order for there to be a copy, there must be an original. In the above photos you see an artist's concept of Ezekiel's "wheel within a wheel". If you research this on Google images, you'll seeing varying yet similar concepts of it. This message came to Ezekiel from God. I find the key point in this being that "the spirit was in the wheels".

Satan had to interpret Ezekiel's dream. In doing so, he came up with his own rendition of a "wheel within a wheel" - hence the LHC. "The spirit is in the wheel" concept is carried over. Nothing original here...

The LHC is a weapon. It is a spirit/energy detector. It will attempt to identify, apprehend, and contain "spirit/s". It will measure varying energy signatures in order to identify the source. Certain energy fields will indicate that which is from God and other energy levels will identify those entangled with Satan. The energies will be different. Satan will need to identify his own in order to control uprisings within his own ranks. His character dictates attributes of greed, power, and control will be key identifiers/markers of those affiliated with him.

The North Area

Upon completion of an acceleration run of particles comprising the contra-rotating beams moving at **0.999999** the speed of light (c), these beams are 'dumped', meaning they are sent to the **North Area** impact point where the particles harmlessly are absorbed along with their attendant energies. Safety systems triggered due to a sensed malfunction, likewise direct the beams to this point. This is where the impacting of Stranglets against Strangelets will occur.

First, a grouping of this quark-gluon condensate will be allowed to escape the Main Ring, also referred to as 'the pipe', and directed toward the North Area impact point.

A second group of Strangelets will be used as the actual explosive required in the separation of the quarks.

https://physicsworld.com/a/quark-gluon-mania-returns-to-cern/

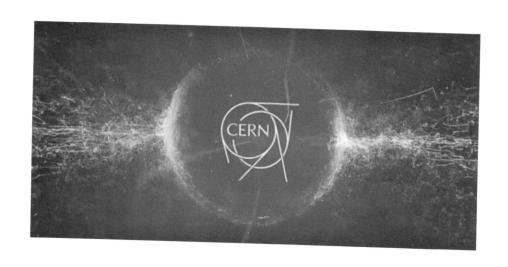

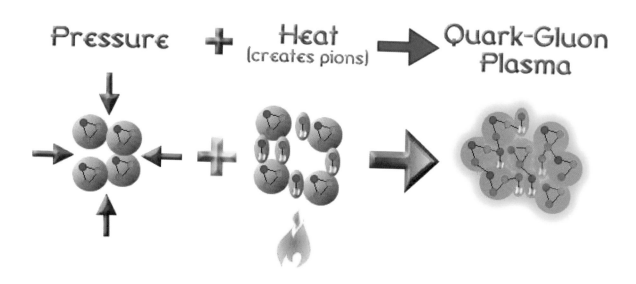

The key to opening the abyss, the "recipe" if you will, is contained in Revelation 9:1. For this is THE key employed in the decrypting/deciphering of an encrypted/ciphered set of instructions. Please refer to past articles within Entangled magazine regarding encryption, Shor's algorithms 2048, 4096, and public key RSA encryption involving D-Wave System's adiabatic quantum computers, designed and employed specifically for this purpose.

The question arises as to just how much energy and power is needed to overcome the Strong Nuclear Force binding the quarks of Strangelets together. Strangelets, by comparison, render nuclear weapons to the level of firecrackers.

Revelation 9:1 KJV

And the fifth angel sounded, and I saw a star fall from heaven unto the earth: and to him was given the key of the bottomless pit.

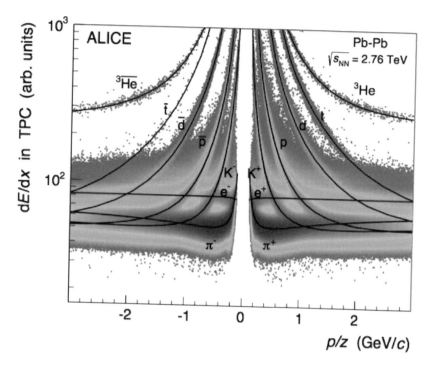

Returning to our graph, imagine at either side of the vertical white area; quarks. They've just been struck by other quarks moving at just below the threshold of the speed of light. Should quarks or any other matter achieve the square of light speed, they would spontaneously convert to energy. Einstein's famous equation:

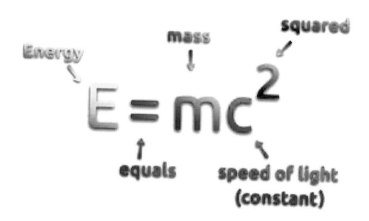

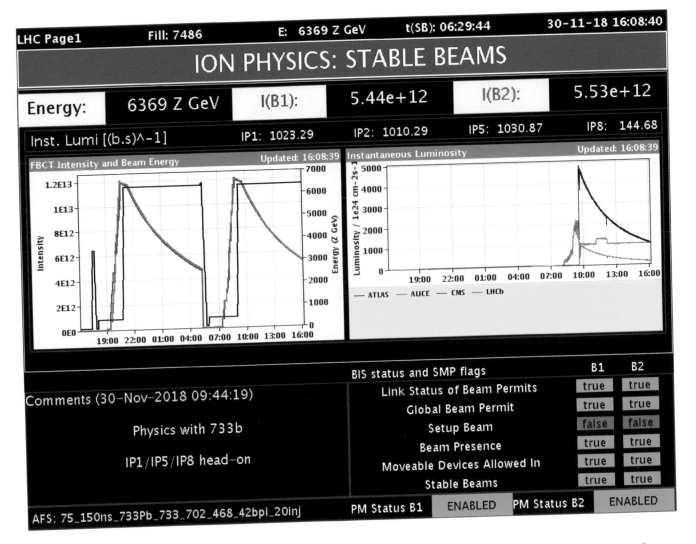

Screen shot of the two beams within the Large Hadron Collider taken November 30, 2018.

I estimate 20 PeV to be the actual power required to overcome and break the Strong Nuclear Force binding quarks and opening the doorway to the abyss.

As seen in this graph, 6369 GeV, or 6.369 TeV per beam (x2) are being realized from the colliding of heavy ions of lead, totalling 12.738 TeV just below the typical level of around 13.5 TeV.

Again, we are moving from Tera (trillion) to Peta (quadrillion) electron volts, defined in several applications as, the mass-energy equivalence, a unit of mass. Also, a unit of momentum and distance and temperature.

D-Wave and CERN

https://www.nature.com/news/quantum-machine-goes-in-search-of-the-higgs-boson-1.22860

This brings us full circle around the Main Ring of the LHC to D-Wave Systems and their 2000Q (2048 qubit chipset) adiabatic quantum computer. The link takes you to an article concerning the use of this computer to re-confirm the discovery of the Higgs boson. Whether this is empirically factual remains debatable. It illustrates a long-standing working relationship between the two organizations.

I have stated that if one seeks to open a 'dimensional doorway' beginning at the quark-quantum scale, you must employ a control mechanism matching this scale. A quantum computer is such a device - a divining device employing geomancy as its core operating system.

"If you think you understand quantum mechanics, you don't understand quantum mechanics." - **Richard Feynman, American Physicist**

This is probably a paraphrase of the quote attributed to earlier Danish Physicist Niels Bohr:

"Anyone who is not shocked by quantum theory has not understood it."

And this time with some variation, similar statements by Feynman and Einstein:

"I couldn't reduce it to the freshman level. That means we really don't understand it." -**Feynman**

"If you can't explain it to a six year old, you don't really understand it."

Feynman lecturing at CERN, 1965

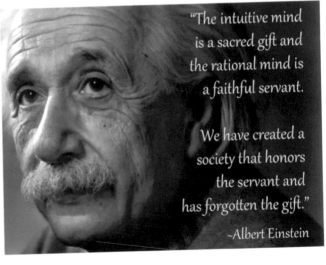

"Mathematicians are only dealing with the structure of reasoning, and they do not really care what they are talking about. They do not even need to know what they are talking about...But the physicist has meaning to all his phrases....In Physics, you have to have an understanding of the connection of words with the real world."

- Richard Feynman in *The Character of Physical Law* (1965),55.

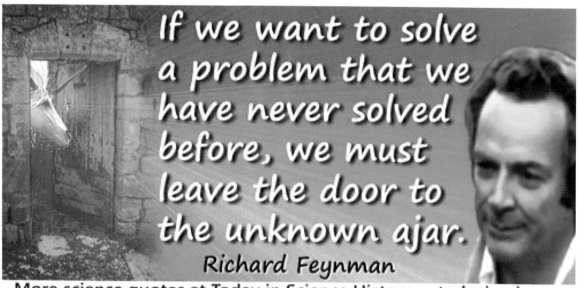

It is the facts that matter, not the proofs. Physics can progress without the proofs, but we can't go on without the facts... if the facts are right, then the proofs are a matter of playing around with the algebra correctly.   Richard Feynman

More science quotes at Today in Science History   todayinsci.com

If we want to solve a problem that we have never solved before, we must leave the door to the unknown ajar. Richard Feynman

More science quotes at Today in Science History   todayinsci.com

As with all quantum-scale particles, each is entangled with another. This, by God's design, is not an isolated system. Therefore, if you affect the spin of one, simultaneously, you affect the other - up spin and down spin. Entanglement is faster-than-the-speed of light communication and effect.

The world, since the first acceleration and collisions of protons, ions of gold, and lead has experienced these changes in the spin of quantum particles *outside* of these colliders.

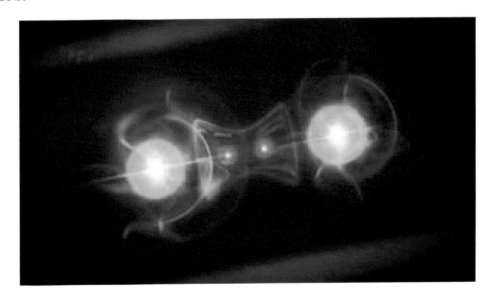

Proponents and prognosticators of quantum computing claim ongoing accessing of near-infinite numbers of 'parallel dimensions', essentially, portals, *without* the use of a particle collider. Quantum computers are inter-dimensional communications devices. They obtain technical information that has resulted in our modern technologies. Technology, a word taken from the Greek "techne" means "science of the craft", or "slight of hand".

This geomantic instrument, a predecessor of today's quantum computer, is from Syria or Egypt, dating to the thirteenth century, was used as an <u>oracular device</u>, a system of magic or the occult.

Through the use of oracular devices, specifically in this examination, quantum computers, in concert with particle colliders, both of which employ exactly the same quantum mechanics only on different scales; <u>entangled particles within our human dimensions continue undergoing alterations in the up and down spin (ex. 0,1,2 integer bosons) of the quantum particles of our reality.</u>

Spin is one of two types of angular momentum in quantum mechanics, the other being orbital angular momentum, much like a spinning gyroscope.

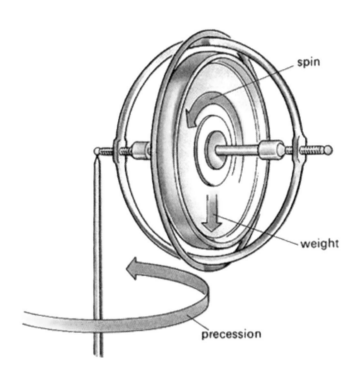

A significant aspect to the operation of quantum computers employing quantum bits, referred to as qubits, is their ability within the chipset to alter the spin of quantum particles, be they electrons or photons. Changing the spin changes a quantum bit from being a zero, to a one and vice versa. Within the world of quantum, a state of superposition of these bits exists. This means a 0 simultaneously can be a 1, and the same for the 1.

What this means: <u>Continually, our physical and spiritual world, the two not being separate, is undergoing changes brought about by the activities of quantum computers and quantum particle colliders.</u>

Inescapably, this remains the case due to the characteristics of quantum mechanics. Specifically, quantum entanglement.

Let me create a scenario for you to relate to:

CERN schedules a collision on Tuesday. D-Wave Systems is standing by to collect the data from this collision. The data they receive from CERN is not limited to simply trajectories, mass, and energies of the collided particles, but D-Wave's role is to look for humanity's resultant actions at the moment of the collision. What CERN wants to know from D-Wave is what effects the quantum particle collisions have upon human behavior. In other words, what happens to humanity at the moment of collisions - how do people respond? Taking into account quantum entanglement of particles, if we **AFFECT** the spin of quantum particles, what is the **EFFECT** and outworking in terms of human behavior?

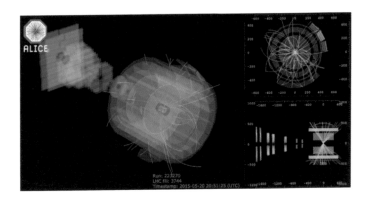

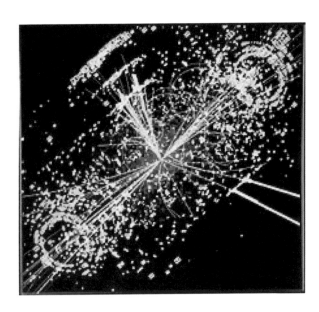

The 4 main detectors, ALICE, ATLAS, CMS and LHCb are where the collisions occur. With varying experiments within each, the quark-gluon condensate Strangelets are being produced.

Yet, due to their inherent design to detect only charged particles (positive or negative), neutral Strangelets only appear briefly, thus described as "winking out of existence". When in fact, and proven with experimental collisions dating back to the 1970s, this plasma condensate, a 'superfluid state of matter', has been produced, sometimes unintentionally, in multiple locations.

Please bear in mind, from a quantum mechanical perspective dealing with measurements of time in picoseconds (one trillionth of a second) and femtoseconds (one quadrillionth of a second), and 13.5 tera electron volts and soon, 20 peta electron volts...consider the ramifications of the following statement presented at the opening of this months magazine made by Dr. Bertolucci in 2009, and at much lower combined beamline energy levels of approximately 7 TeV...

When dealing with quantum timescales, wherein time is for all practical purposes irrelevant; a time frame of between 10 and 26 seconds may as well be an *eternity*. His 2009 statement as the former Director for Research and Scientific Computing at CERN pales in comparison even to today's capabilities of the Large Hadron Collider. Then, we must consider their planned-for upgrades of the largest, most complex, most expensive machine ever built in recorded human history. Surpassing even the Tower of Babel.

Dr. Bertolucci later confirmed that yes indeed, there would be an "open door", but that even with the power of the LHC at his disposal (in 2009) he would only be able to hold it open *"a very tiny lapse of time, 10-26 seconds, [but] during that infinitesimal amount of time we would be able to peer into this open door, either by getting something out of it or sending something into it."*

Working with these short amounts of time and large energy scales requires the use of quantum and classical computers. It should be noted, computers originated from geomancy.

The October 2017 issue of Entangled magazine presented a short history of geomancy, its direct relationship to modern computing, and its use by Jesus in the Temple.

Your Unique Source For Leading-Edge Insights Into The Hidden Aspects of Science and Biblical Scripture

Entangled

VOLUME-5 October 2017

Science and technology news:
AnthonyPatch.com

ISBN 978-1-9739-1746-5

9 781973 917465

John 8:2-11 KJV

(highlighting verse 6)

6 This they said, tempting him, that they might have to accuse him. But Jesus stooped down, and with his finger wrote on the ground, as though he heard them not.

Jeremiah 17:13 King James Version (KJV)

13 O Lord, the hope of Israel, all that forsake thee shall be ashamed, and they that depart from me shall be written in the earth, because they have forsaken the Lord, the fountain of living waters.

Please refer to the extensive biblical references and commentary provided within the pages of the November 2017 issue of Entangled magazine specific to Jesus communicating by symbolism directly to the Scribes and Pharisees, "learned men", who were taught the art of divination, specifically in this case using geomantic symbols.

When Jesus wrote the geomantic symbols in the sand, these learned men, Scribes and Pharisees, knew the message he was conveying. They were schooled in the interpretations. He affirmed by using the figures: Via, Populus, Puella and Carcer. In choosing Via, he stated that he was the way, the truth, and the life. He used Populus to show that he would sway the crowd to do what was right and calm it down quenching their blood thirst affirming his position as judge. Puella represented the young woman thrown at his feet. By choosing Carcer, he stated that they'd be imprisoned for their acts against God. It was after this incident, and their interpreting of Jesus' four geomantic figures that they increased their efforts to ultimately imprison and kill him. They knew then that Jesus knew their private language and it infuriated them.

The image of Jesus writing in the sand represents His turning the tables on the Scribes and Pharisees in the Temple; and taking their system of divination and that which they were schooled in and using it against them.

Similarly, Daniel was trained in the art of divination, though he remained true to God in his heart. God strategically positioned him against the Babylonian magicians to decipher King Nebuchadnezzar's dream (Daniel 2).

We know the teachers of the law at the time were trained in the "forbidden arts" and hidden knowledge. This black magic was taught in the days of Daniel. God used that language to show Daniel the King's dream.

High priests of the Levitical Priesthood cast lots and used talismans such as the Urim and Thummim stones to arrive at a judgment. It is why God told Daniel to "close the book" and not look into it any further until the end of days.

Daniel 12:4 KJV

But thou, O Daniel, shut up the words, and seal the book, even to the time of the end: many shall run to and fro, and knowledge shall be increased.

URIM: *oo-reem'*

lights; *Urim*, the oracular
brilliancy of the figures
in the high priests breastplate
as an emblem of *complete* LIGHT.

THUMMIM: *toom-meem'*

perfections, that is, one of
the epithets of the objects
in the high priests breastplate
as an emblem of *complete* TRUTH.

Sample Bible verses on the Urim and Thummim:

1 Samuel 28:5-6

Exodus 28:30

Nehemiah 7:63-65

Numbers 27:21

Leviticus 8:7-8

Deuteronomy 33:8

Ezra 2:62-63

Abraham 3:1-2

Abraham 3:4

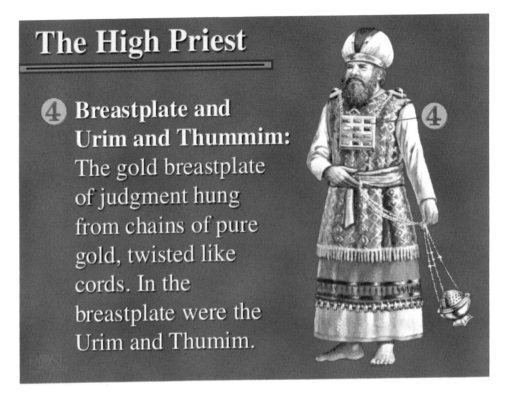

The Standard Model of Physics

https://youtu.be/V0KjXsGRvoA

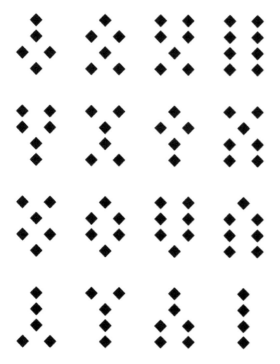

Your Unique Source For Leading-Edge Insights Into The Hidden Aspects
Of Science & Biblical Scripture

Entangled
MAGAZINE
VOLUME-12   MAY 2018

❋ ANTICHRSIT INFRASTRUCTURE   ❋ D-WAVE'S QUADRANT
❋ GEOMANTIC TECH   ❋ QUANTUM SUPREMACY

The 16 figures of Geomancy of which there are 65,536 possible combinations, equal to the number of tubulin dimers in one microtubule of a human neuron.

May, 2018 Issue

We continue making marks in the sand, the silica and silicon of sand.

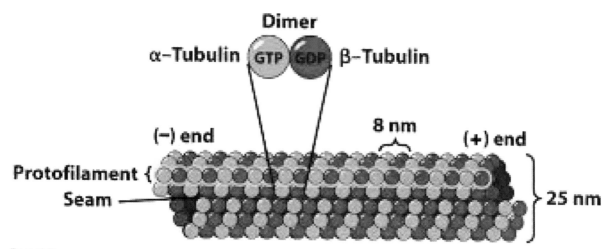

Figure 19-16
Molecular Cell Biology, Sixth Edition
© 2008 W.H. Freeman and Company

Please note the image results when searching the term 'oracular device':

https://images.search.yahoo.com/search/images;_ylt=AwrJ7JUXagFcsSQA7NZXNyoA;_ylu
=X3oDMTB0N2Noc21lBGNvbG8DYmYxBHBvcwMxBHZ0aWQDBHNlYwNwaXZ?p
=oracular+device.&fr2=piv-web&fr=mcafee

 Oracular Device

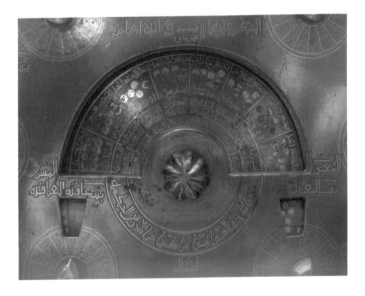

The January issue of Entangled magazine will provide further details of Ezekiel's wheel within a wheel and what it means for us. We will look at the direct correlation between the use of the detectors in measuring the Luminosity arising from collisions of physical matter and their detection of spirits.

Included are the topics of time & seasons based upon the true definition of 'quantum'. What is meant by dimensions according to scientists and the lay person. Then, a presentation of Jesus as the true Singularity.

I hope you will find the information interesting and beneficial, and thank you for subscribing to Entangled magazine.

Anthony Patch

Founder, Entangled Magazine

The Adventure Begins

The bond of sisters, the story of friends, the transgression of innocence, and

the empowerment of hope...

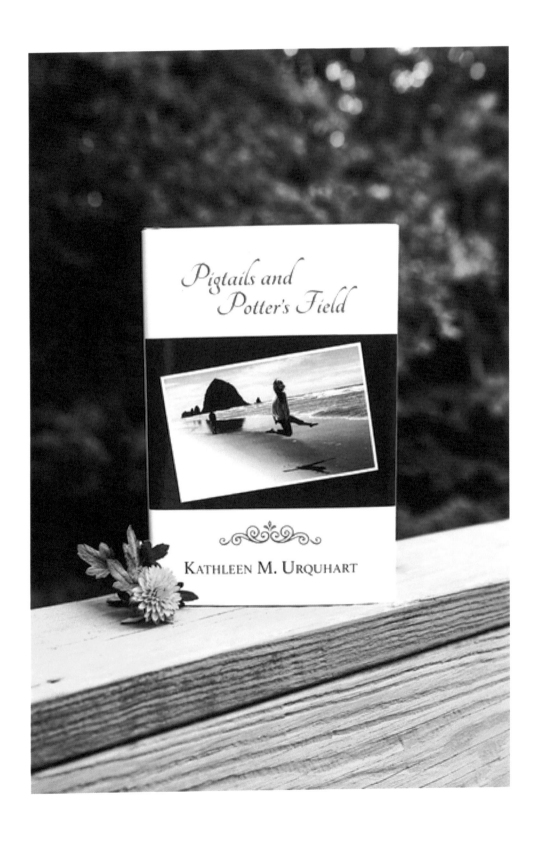

Please consider my wife's new novel; *Pigtails and Potter's Field.*

For more research and reporting, please join me on Patreon:

https://www.patreon.com/anthonypatch

You will receive access to private Livestreams on Wednesdays and Thursdays, as well as broadcast archives, including those heard Friday's on my new BlogTalkRadio program:

http://www.blogtalkradio.com/anthonypatchshow

And, archives of all past Webinars and exclusive articles posted only for Patrons.

Thank you and God Bless You and Yours,

Anthony Patch

Please refer to my second novel in the series (see Covert Catastrophe; 2013) 2048: Diamonds in the Rough (2014) for more on this entanglement, as well as extensive coverage provided within past issues of Entangled magazine.

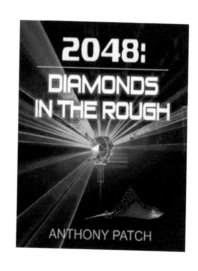